FLORA OF TROPICAL EAST AFRICA

VELLOZIACEAE

Lyman B. Smith and Edward S. Ayensu *

Perennial herbs or shrubs; indument highly diverse, often not strictly epidermal; ** stems mostly elongate and branched, covered with persistent leaf-sheaths, fibrous. Leaves (young, complete) in a fascicle at the end of the stem or of each branch; sheaths densely imbricate, cylindric with a V-shaped sinus opposite the blade; blades narrow and grass-like, persistent and gradually decomposing with age or deciduous by a symmetrical transverse line. Peduncles at the end of the stem or of each branch, 1-flowered. Flowers perfect or rarely functionally unisexual (*Barbaceniopsis*). Perianth-tube equalling to greatly exceeding the ovary and adnate to it; tepals 6 in 2 series but usually almost identical. Stamens 6 or numerous (*Vellozia* in part); filaments terete and slender or variously flattened and forming a corona-like ring; anthers linear, introrse, opening by longitudinal slits. Ovary 3-locular, inferior; style slender, elongate; stigmas linear and vertical or orbicular and horizontal from the apex of the style (*Vellozia*); ovules numerous in many rows on stalked placentas. Fruit a tardily and irregularly dehiscent capsule. Seeds numerous; embryo small in a copious endosperm.

A family of three or four genera from the New World including the type genus *Vellozia* Vand., one from Africa, Madagascar, and Arabia, and one from South Africa.

XEROPHYTA

Juss., Gen.: 50 (1789); Lam., Tabl. Encycl. 1: 372 (1792); Bak. in J.B. 13: 231 (1875); Perrier in Fl. Madag. 42, Velloz.: 2 (1950); L.B. Smith & Ayensu in K.B. 29: 184 (1974)

Small to large perennial herbs or shrubs. Leaf-blades deciduous along a regular transverse line or rarely persistent (*X. schlechteri* (Bak.) N. Menezes). Perianth-tube slightly exceeding the ovary; tepals white, blue, mauve, or yellow. Filaments flattened and almost wholly adnate to the tepals, triangular, sometimes auricled near apex. Stigmas linear, vertical, about the same length or much longer than the naked style-base.

A genus of twenty-nine species, with the type, *X. pinifolia* Lam., from Madagascar along with two others, one from Arabia, and the remainder from tropical Africa southwards.

We recognise in K.B. 29: 184 (1974) three sections based on leaf-anatomy. The two sections occurring in E. Africa are defined as follows.

Sect. *Barbacenioides* L. B. Smith & Ayensu. Leaf-anatomy of *Barbacenia* type: adaxial sclerenchyma inverted Y-shaped, the abaxial Y-shaped or 3-pronged. Mesophyll isolateral, not differentiated into palisade and spongy tissues. Species 1, 2, 5–7.

Sect. *Vellozioides* L. B. Smith & Ayensu. Leaf-anatomy of *Vellozia* type (fig. 1) adaxial sclerenchyma inverted V-shaped, abaxial U- or Y-shaped. Mesophyll dorsiventral, with distinct palisade and spongy tissues. Species 3, 4, 8.

Fuller discussion is given in K.B. 23: 315–335 (1969).

* Smithsonian Institution, Washington, D.C., U.S.A.
** Greves in J.B. 59: 274 (1921).

Ovary glabrous; tepals 40–50 mm. long; *Barbacenia*
  type leaf-anatomy .   .   .   .   .   1. *X. goetzei*
Ovary covered with trichomes:
  Trichomes of the ovary simple and attenuate to
    stellate, non-glandular:
    Ovary covered with very short hairs or setae;
      Barbacenia type leaf-anatomy .   .   .   2. *X. equisetoides*
    Ovary covered with broad-based trichomes:
      Trichomes of the ovary predominantly simple,
        terete, subulate or setose with a tumid base;
        Vellozia type leaf-anatomy:
        Leaf-blades glabrous; stems mostly short and
          simple; ovary subglobose .   .   .   3. *X. schnizleinia*
        Leaf-blades pubescent at least toward base;
          stems mostly tall and branched; ovary
          narrowly ovoid to subcylindric .   .   4. *X. simulans*
      Trichomes of the ovary predominantly forked
        or stellate:
        Leaf-blades glabrous or laxly vestite beneath
          with subterete trichomes:
          Ovary-trichomes predominantly forked
            with long slender rays; leaf-blades
            laxly vestite beneath with ± plumose
            trichomes; *Vellozia* type leaf-anatomy   4. *X. simulans*
          Ovary-trichomes broadly stellate with short
            rays; leaf-blades glabrous beneath or
            with linear trichomes; *Barbacenia*
            type leaf-anatomy .   .   .   .   5. *X. suaveolens*
        Leaf-blades covered beneath with flat linear-
          lanceolate trichomes; ovary-trichomes
          orbicular with short broad rays; *Barba-*
          cenia type leaf-anatomy .   .   .   6. *X. nutans*
  Trichomes of the ovary glandular:
    Stem tall and branching; tepals much more than
      15 mm. long, acuminate to apiculate:
      Leaf-blades pubescent at least toward base:
        Barbacenia type leaf-anatomy .   .   .   7. *X. spekei*
      Leaf-blades scabrous above, otherwise glabrous;
        Vellozia type leaf-anatomy .   .   .   8. *X. scabrida*
    Stem very short with leaves subrosulate; tepals
      not more than 15 mm. long, the outer
      aristate; *Vellozia* type leaf-anatomy .   .   *X. humilis**

1. **X. goetzei** (*Harms*) *L. B. Smith & Ayensu* in K.B. 29: 188 (1974).
Type: Tanzania, Iringa District, Little Ruaha R., *Goetze* 426 (B, holo.!, K, iso.)

Stems erect, stout, branched, 1–2 m. high; branches (including the
leaf-sheaths) cylindric, slightly conical at apex, 8–10 mm. thick. Leaf-
sheaths convex, obscurely carinate, dark castaneous, densely pubescent
beneath toward apex, remaining straight after abscission and dividing
tardily; blades linear, filiform-attenuate, 14–23 cm. long, 4–8 mm. wide,
subsericeous-pubescent throughout at first, very obscurely serrulate toward

* *X. humilis* (Bak.) Th. Dur. & Schinz is not yet known for the Flora area, but
recorded from S. Sudan, between Lake Rudolf and Gondokoro (*Donaldson Smith*), and
likely to occur, the main part of its range being south of the Flora area in Mozambique,
Zambia, Rhodesia, South Africa and South West Africa.

apex; anatomy of the *Barbacenia* type. Peduncles 4–7 cm. long, glabrous. Perianth pale blue; tepals linear-lanceolate, acuminate, 4–5 cm. long, glabrous. Ovary narrowly ellipsoid, glabrous.

TANZANIA. Iringa District: Little Ruaha R., 4 Jan. 1899, *Goetze* 426!
DISTR. T7; not known elsewhere
HAB. Gregarious, among rocks, dry sunny mountain slopes; 600 m.

SYN. *Barbacenia goetzei* Harms in E.J. 28: 363 (1900)

2. **X. equisetoides** *Bak.* in J.B. 13: 233 (1875). Type: Malawi, Zomba and E. end of Lake Chilwa [Shirwa], *Meller* (K, holo.!)

Stems to 2·6 m. high with few branches. Leaf-sheaths dark castaneous, short-pubescent or appressed-setose at apex; blades linear, attenuate, 4–70 cm. long, 1·5–8 mm. wide; anatomy of the *Barbacenia* type. Peduncles 1–4 at the apex of the branch, 2·5–9 cm. long, very slender and flexuous, densely pubescent to almost wholly glabrous. Perianth white to pale mauve, or pink, fragrant; tepals nearly uniform, linear-lanceolate or oblong, acute, appearing acuminate by the inrolling of the apex, 20–35 mm. long. Filaments adnate to the tepals, triangular; anthers linear, ± half as long as the tepals. Ovary ellipsoid, 6–12 mm. long, covered with appressed subulate brownish setae; style-base much shorter than the stigma.

KEY TO INFRASPECIFIC VARIANTS

Branches (including leaf-sheaths) very short and stout,
  ± as long as thick; leaf-sheaths soon splitting, 4–5
  cm. long, very slightly exserted; leaf-blades
  soft-hairy on both sides . . . . . var. **trichophylla**
Branches (including leaf-sheaths) long and slender,
  much longer than thick; leaf-sheaths not
  splitting, 3 cm. long, much exserted:
  Blades glabrous or with only marginal trichomes . var. **pauciramosa**
  Blades soft-hairy . . . . var. **pubescens**

var. **trichophylla** (*Bak.*) *L. B. Smith & Ayensu* in K.B. 29: 192 (1974). Type: Malawi, without precise locality, *Buchanan* 854 (K, lecto.!, BM, US!, isolecto.)

Branches (including leaf-sheaths) very short and stout, ± as long as thick. Leaf-sheaths soon splitting, 4–5 cm. long, very slightly exserted; blades soft-hairy on both surfaces.

TANZANIA. Masasi District: Tunduru–Masasi road, near Nangua, Nov. 1951, *Eggeling* 6391!
DISTR. T8; Zaire, Zambia, Malawi, Mozambique and Rhodesia
HAB. Rock outcrops; 450 m.

SYN. *Vellozia equisetoides* Bak. var. *trichophylla* Bak. in F.T.A. 7: 411 (1898)
     [*V. equisetoides* sensu W. Watson in Gard. Chron., ser. 3, 34: 425, fig. 167 (1903), *non* Bak.]
     V. trichophylla (Bak.) Hemsl. in Bot. Mag. 130, t. 7962 (1904); Greves in J.B. 59: 281 (1921)
     ? *Xerophyta barbarae* Duvign. & Dewit in B.S.B.B. 96: 145, photo. 16, 17, fig. 10 (1963)
     X. trichophylla (Bak.) N. Menezes in Ciência e Cultura 23: 422 (1971)

var. **pauciramosa** *L. B. Smith & Ayensu* in K.B. 29: 192, fig. 2 (1974). Type: Zambia, Western Province, Mwinilunga District, near Zambesi R., 6·5 km. N. of Kalene Hill Mission, *F. White* 3369 (BM, holo.!, BR!, K, iso.)

Branches (including leaf-sheaths) long and slender, much longer than thick. Leaf-sheaths not splitting, 3 cm. long, much exserted; blades glabrous or with only marginal trichomes.

TANZANIA. Mbeya District: Sante on Niamba [Yamba] R., *Goetze* 1409!
DISTR. T7; Zaire, Zambia, Rhodesia, South West Africa
HAB. *Brachystegia* woodland, rock outcrops; 1200 m.

SYN. *Barbacenia wentzeliana* Harms in E.J. 30: 277 (1902). Type: Tanzania,
      Mbeya District, Sante on Niamba [Yamba] R., *Goetze* 1409 (B, holo. !, BM, iso. !)
      Vellozia wentzeliana (Harms) Greves in J.B. 59: 281 (1921), excl. *Rand* 155
      (Transvaal); T.T.C.L.: 629 (1949)
      Barbacenia naegelsbachii E. Holzh. in Mitt. Bot. Staats. München 1, Heft 8:
      334 (1953). Type: South West Africa, northern section, Kamanjab,
      Naegelsbach (M, holo. !)

var. **pubescens** *L. B. Smith & Ayensu* in K.B. 29: 194 (1974). Type: Rhodesia, Mazoe
District, near Tstatsi R., about 22 km. N. of Concession, *Leach* 11283 (K, holo. !, P, iso. !)

Branches and leaf-sheaths as in var. *pauciramosa*. Blades pubescent.

TANZANIA. Ufipa District: Lake Rukwa, Milepa, 10 Dec. 1935, *Michelmore* 1152 !
DISTR. **T4**; Malawi, Zambia
HAB. Rock outcrops, *Brachystegia* woodland; 600 m.

DISTR. (of species as a whole). **T4, 7, 8**; Zaire, Zambia, Malawi, Mozambique, Rhodesia,
      South West Africa and South Africa (Transvaal)

3. **X. schnizleinia** (*Hochst.*) *Bak.* in J.B. 13: 235 (1875), as "*schnitz-
leinia* ", & in F.T.A. 7: 409 (1898); Th. Dur. & Schinz, Consp. Fl. Afr. 5:
272 (1895). Type: Ethiopia, Semien, Sabra, *Schimper* 1365 (BM !, K, iso.)

Plant very variable in size of parts. Stems simple or rarely few-branched
from the base, 1–6(–50) cm. high, (including the leaf-sheaths) terete, subcylindric or somewhat thickened at base, (5–)10–15 mm. thick. Leaf-sheaths
convex and ecarinate, dull brown, remaining nearly straight, after abscission
dividing into coarse fibres with few or no fine cross-fibres; blades linear,
filiform-attenuate, (3–)10–30 cm. long, 2–6 mm. wide, setose-serrate near
apex on the margins and on the underside of the midnerve, otherwise entire
and glabrous; anatomy of the *Vellozia* type. Peduncles 1–3 at the apex of
the stem, 1–28 cm. long, setose near apex. Perianth white or flushed with
mauve or pink; outer and inner tepals nearly identical, linear-lanceolate,
acute, (5–)18–40 mm. long, setose at base. Filaments subtriangular, to
3 mm. long, almost wholly adnate to the tepals; anthers linear-lanceolate,
attached 2 mm. above the bilobed base, up to 15 mm. long. Ovary subglobose with a short epigynous tube, covered with broad-based subulate
stiff trichomes; stigmas linear, apical. Fruit globose, up to 10 mm. wide.

UGANDA. Karamoja District: Turkana Escarpment, Lutyen Pass, 5 Mar. 1936,
      Michelmore 1242 !
KENYA. Northern Frontier Province: Dandu, 14 May 1952, *Gillett* 13195 ! & Mathews
      Range, Olkanto, 10–12 Dec. 1944, *J. Adamson* 40 in *Bally* 4341 ! & Mathews Peak
      [Ol Doinyo Lengiyo], 20 Dec. 1958, *Newbould* 3270 ! & 3510 !
DISTR. **U1**; **K1, 2**; Ethiopia, Somali Republic, Nigeria
HAB. Rocks or sunk in soil; 650–1800 m.

SYN. *Hypoxis schnizleinia* Hochst. in Flora 27: 31 (1844)
      Vellozia schnizleinia (Hochst.) Martelli, Fl. Bogos.: 82 (1886), as " *schnitzleinia* ";
      Bak. in F.T.A. 7: 409 (1898); Greves in J.B. 59: 280 (1921)
      Barbacenia schnizleinia (Hochst.) Pax in Engl., Hochgebirgsfl. Trop. Afr.:
      171 (1892), as " *schnizleiniana* "
      B. hildebrandtii Pax in Engl., Hochgebirgsfl. Trop. Afr.: 171 (1892). Type:
      Somali Republic (N.), Serrut Mts., Meid, *Hildebrandt* 1466 (B, holo. !)
      Vellozia schnizleinia (Hochst.) Martelli var. *somalensis* Terracc. in Boll. Soc. Bot.
      Ital. 1892: 425 (1892). Type: Somali Republic, Gerar-Amaden, *Candeo &
      Baudi di Vesme* (?FI, holo.)
      V. hildebrandtii (Pax) Bak. in F.T.A.7: 409 (1898); Greves in J.B. 59: 280 (1921)
      V. somalensis (Terracc.) Chiov. in Ann. Bot. Roma 9: 141 (1911)
      V. schnizleinia (Hochst.) Martelli var. *occidentalis* Milne-Redh. in K.B. 3: 381
      (1951); Hepper in F.W.T.A., ed. 2, 3: 174, fig. 378 (1968). Type: Nigeria,
      Zaria District, Anara Forest Reserve, *Keay* in *F.H.I.* 22903 (K, holo. !)
      Xerophyta somalensis (Terracc.) N. Menezes in Ciência e Cultura 23: 422 (1971)

NOTE. Terracciano's var. *somalensis* has nothing but small size to distinguish it and
      subsequent collections show such intergrading as to make it untenable. Hepper's
      illustration of Milne-Redhead's variety *occidentalis* shows two plants growing together,
      one with the elongate stem of the proposed variety, the other in no way different
      from the typical species.

FIG. 1. *XEROPHYTA SIMULANS*—**1,** flowering branch, × ½; **2,** trichomes of leaf × 25; **3,** trichomes of ovary, × 25; **4,** lateral section of upper part of flower, × 2; **5,** transverse section of leaf-blade, × 62; **6,** same, detail, × 124. All from *A. S. Thomas* 4046.

4. **X. simulans** *L. B. Smith & Ayensu* in K.B. 29: 189, fig. 1 (1974). Type: Uganda, Acholi District, Gulu, Kilak, *A. S. Thomas* 4046 (K, holo. !)

Stem to 2·4 m. high, (including leaf-sheaths) terete; ultimate branches erect, very short, 10 mm. thick. Leaf-sheaths ± 4 cm. long, dark castaneous, quickly disintegrating after abscission; blades linear, attenuate, to 30 cm. long and 7 mm. wide, sparsely pubescent at least near base with straight, white, ± plumose hairs; anatomy of the *Vellozia* type. Peduncles 1–4 at the apex of each branch, 1–5 cm. long, crisped-setose toward apex. Perianth white or pale mauve, almond-scented; tepals nearly uniform, linear-lanceolate, acuminate, 20–35 mm. long. Filaments triangular, almost wholly adnate to the tepals; anthers linear, ± half as long as the tepals, attached 3 mm. above the base which extends below the filament. Ovary narrowly ovoid to subcylindric, 6–10 mm. long, slightly trigonous, its trichomes very dense, typically simple and ovoid with apiculate or subulate apex but varying to few-forked; stigmas linear, vertical, apical, ± as long as the style-base. Capsule ellipsoid, 20 mm. long. Seeds suboblong, coarsely reticulate. Fig. 1, p. 5.

UGANDA. Acholi District: Chua, Adilang, July 1937, *Eggeling* 3349 ! & Kilak, *A. S. Thomas* 4046 !; Karamoja District: Kotido, 2 June 1940, *A. S. Thomas* 3677 !
TANZANIA. Kigoma District: without precise locality, *Grant* ! & 32–48 km. S. of Uvinza, Buyenzi, Feb. 1956, *Procter* 420 !; Mpanda District: Kala, 11 Nov. 1963, *Carmichael* 1012 !
DISTR. U1; T4; Sudan, Zambia, Rhodesia
HAB. Dry rocks, short turf over granite outcrops; 900–1200 m.

5. **X. suaveolens** *(Greves) N. Menezes* in Ciência e Cultura 23: 422 (1971). Type: Rhodesia, Mazoe District, Bernheim Hill, *Eyles* 439 (BM, lecto. !)

Stem 12–18 dm. high, (including leaf-sheaths) terete, to 8–15 cm. thick at base, ultimate branches 10–15 mm. thick. Leaf-sheaths (incompletely known) over 25 mm. long, blackish brown, quickly disintegrating after abscission; blades linear, filiform-attenuate, to 65 cm. long and 12 mm. wide, the margins inconspicuous, yellowish, entire, the trichomes simple, subterete; anatomy of the *Barbacenia* type. Peduncles 1–4 at the apex of each branch, 6–13 cm. long, covered with flat forked or stellate trichomes toward apex. Perianth white to pale lavender, sweet scented; outer and inner tepals nearly identical, narrowly lanceolate, acuminate, 35–60 mm. long. Filaments subquadrate, almost wholly adnate to the tepals; anthers linear, ± half as long as the tepals. Ovary ellipsoid, 8–16 mm. long, covered with flat erect forked or stellate trichomes; stigmas linear, vertical, apical, much longer than the style-base.

var. **suaveolens**

Leaf-blades glabrous and glaucous beneath, scabrous above.

TANZANIA. Songea District: Matagoro Hills just S. of Songea, 3 Feb. 1956, *Milne-Redhead & Taylor* 8592 !
DISTR. T8; Zaire, Zambia, Mozambique, Rhodesia, Botswana
HAB. Bare exposed rocks; 1450 m.

SYN. *Vellozia suaveolens* Greves in J.B. 59: 282 (1921)

var. **vestita** *L. B. Smith & Ayensu* in K.B. 29: 194 (1974). Type: Zambia, 48 km. E. of Mbala [Abercorn] on Tunduma road, *Napper* 1161 (EA, holo. !, K, iso. !)

Leaf-blades from slightly to much longer vestite than scabrous.

TANZANIA. Without locality,* *Busse* 646 !; Songea District: Matagoro, Nov. 1951, *Eggeling* 6363!
DISTR. T8; Zambia
HAB. Rock, cliff in *Brachystegia* woodland; 1200 m.

* Probably Kilwa District, Donde

6. **X. nutans** *L. B. Smith* & *Ayensu* in K.B. 29: 195, fig. 3 (1974). Type: Tanzania, Masasi District, *Milne-Redhead* & *Taylor* 7687 (EA, holo. !, BR, K, iso. !)

Stem up to 4·5 dm. high; main branches up to 3 cm. thick; flowering branches very short and covered by the leaf-sheaths. Leaf-sheaths blackish brown with grey apices, quickly splitting after abscission; blades linear, filiform-attenuate, up to 30 cm. long and 4 mm. wide, covered beneath with flat linear-lanceolate grey trichomes, soon glabrous above; anatomy of the *Barbacenia* type. Peduncles 3 at the apex of each small branch, 7–9 cm. long, sublaxly stellate-pubescent. Flowers nodding. Tepals nearly uniform, linear-lanceolate, acute, 30 mm. long, pale lilac. Filaments triangular, almost wholly adnate to the tepals; anthers linear, ± half as long as the tepals. Ovary ellipsoid, 8–10 mm. long, covered with orbicular short-rayed trichomes; stigmas linear, apical, much longer than the style-base.

TANZANIA. Masasi District: about 33·5 km. E. of R. Lumesule [Lemesule], 17 Dec. 1955, *Milne-Redhead* & *Taylor* 7687 !
DISTR. T8; known only from the type-gathering
HAB. On smooth exposed rocky slab; 510 m.

7. **X. spekei** *Bak.* in J.B. 13: 234 (1875); Th. Dur. & Schinz, Consp. Fl. Afr. 5: 272 (1895). Type: Tanzania, Tabora District, Boss Rock, *Grant* 628 (K, holo. !)

Plant very variable, to 2(–5) m. high; trunk 10–13 cm. thick. Stems much branched; branches (including leaf-sheaths) subcylindric, slightly tapered towards apex, 6–12 mm. thick. Leaf-sheaths convex, obtusely carinate and typically with the midnerve enlarged into a broad subapical smooth umbo, dark castaneous, lustrous at first, remaining straight or ± recurving after abscission, dividing tardily if at all, densely white strigose towards apex; blades linear, filiform-attenuate, 7–30(–70) cm. long, 3–12 mm. wide, pilose beneath, soon glabrous above, setose-serrate near apex on the margins and underside of the midnerve; anatomy of the *Barbacenia* type. Peduncles 1–3 at the apex of the stem, 2–8 cm. long, laxly glandular. Perianth white to pale blue-lilac or mauve; tepals nearly uniform, linear-lanceolate, acuminate to rounded and retuse-apiculate, 20–35(–50) mm. long, glandular at base. Filaments broadly triangular, almost wholly adnate to the tepals; anthers linear, up to 13(–20) mm. long. Ovary ellipsoid with a short epigynous tube, 5–10 mm. long, subdensely vestite with subsessile glands; stigmas linear, subapical. Fruit subglobose, up to 15 mm. long.

KENYA. Northern Frontier Province: Dandu, Mar.–Apr. 1952, *Gillett* 12635 !;
    Machakos District: between Machakos turn-off and Kibwezi, 30 Nov. 1959, *Greenway* 9611 !; Teita District: Gimba Hill, *Gardner* in F. D. 2924 !
TANZANIA. Handeni District: Mzinga Hill, 6 Sept. 1933, B. D. *Burtt* 4820 !; Kilosa
    District: Mamboya, Jan. 1931, *Haarer* 1971A ! & Berega–Mlali on Mpwapwa road, 10 Dec. 1935, B. D. *Burtt* 5426 !
DISTR. K1, 4, 6, 7; T2–4, 6, 7; Zambia
HAB. Rock outcrops; 450–1900 m.

SYN. *Barbacenia tomentosa* Pax in E.J. 15: 144 (1892). Type: Kenya, Machakos
    District, Athi [Asi] R., *Fischer* 585 (B, holo. !, K, iso.)
    Vellozia spekei (Bak.) Jackson in Ind. Kew.: 1173 (1895), erroneously attributed to Baker; T.T.C.L.: 629 (1949)
    V. *aequatorialis* Rendle in J.L.S. 30: 409 (1895); Greves in J.B. 59: 283 (1921); T.T.C.L.: 628 (1949). Type: Tanzania, between Zanzibar and Uyui, W. E. *Taylor* (BM, holo.)

Xerophyta tomentosa (Pax) Th. Dur. & Schinz, Consp. Fl. Afr. 5:   272 (1895)
Barbacenia aequatorialis (Rendle) Harms in P.O.A. C: 146 (1895)
B. spekei (Bak.) Harms in P.O.A. C: 146 (1895)
Vellozia tomentosa (Pax) Bak. in F.T.A. 7:   412 (1898);   Greves in J.B. 59:
    283 (1921);   T.T.C.L.:   629 (1949)
Barbacenia helenae Buscal. & Muschl. in E. J. 49:   462 (1913);   Piscicelli, Nella
    Regione dei Laghi Equatoriali: 464, fig. (1913). Type: Kenya, Masai District,
    Uaso Nyiro [Guasso Nyiro], *Aosta* 1601 (B, holo. †, identification by description
    and illustration)
Xerophyta aequatorialis (Rendle) N. Menezes in Ciência e Cultura 23:   422 (1971)

NOTE.   A robust form with stems up to 5 m. tall and large flowers (tepals 4–5 cm. long)
occurs in Tanzania, principally around the Usambara Mts.

8. **X. scabrida** (*Pax*) *Th. Dur. & Schinz*, Consp. Fl. Afr. 5:   271 (1895).
Type: Angola, Quango, *Pogge* 423 (B, holo. !)

Stem (unknown in type) 60–80 cm. high, (including leaf-sheaths) terete,
8–14 mm. thick near apex. Leaf-sheaths (unknown in type) to 3 cm. long,
remaining a while after abscission or the extreme apex partially dividing to
form a fine net;   blades linear, attenuate, 15–20 cm. long, 5–7 mm. wide,
± scabrous on the upper side, marginal nerves ± twice as wide as the others,
yellow, finely denticulate;   anatomy of the *Vellozia* type.   Peduncles 3–4
at the apices of the stem, 7–10 cm. long, glandular especially toward apex.
Perianth white to pale mauve;   outer and inner tepals nearly identical,
lanceolate, acute, 25–35 mm. long, glandular, especially the outer.   Filaments
subtriangular, almost wholly adnate to the tepals;   anthers linear, ± two-
thirds as long as the tepals.   Ovary turbinate to ellipsoid, 7–10 mm. long,
densely dark-glandular;   stigmas linear, vertical, apical, much longer than
the style-base.

TANZANIA.   Masai District: 7 km. Kibaya–Kondoa, 16 Jan. 1965, *Leippert* 5463 !;
    N. Mpwapwa, 11 Dec. 1935, *Hornby* 729 !;   Masasi District: Tunduru–Masasi road,
    Nangua, Nov. 1951, *Eggeling* 6390 !
DISTR.   **T**2, 5, 6, 8;   Zambia, Angola
HAB.   Rock outcrops;   450–1900 m.

SYN.   *Barbacenia scabrida* Pax in E.J. 15:   144 (1892)
       Vellozia scabrida (Pax) Bak. in F.T.A. 7:   410 (1898)

INDEX TO VELLOZIACEAE